Do They Have Telephones Up In Heaven?

R. R. Pravin

authorHOUSE®

AuthorHouse™
1663 Liberty Drive
Bloomington, IN 47403
www.authorhouse.com
Phone: 1 (800) 839-8640

Published by AuthorHouse 07/10/2017

ISBN: 978-1-5246-9985-7 (sc)
ISBN: 978-1-5246-9984-0 (e)

Print information available on the last page.

Table of Contents

Foreword

Dr RR Pravin is currently a Year 1 Paediatric Resident at KK Women's & Children's Hospital (KKH) during which he had an attachment to the Paediatric Oncology ward and showed how truly dedicated and professional he was, even as a junior doctor, in looking after sick children with cancer.

Right from the beginning of his medical student days in the National University of Singapore, he assumed a variety of positions in Project Rudolph where a team of medical students would sing carols yearly on Christmas Eve to KKH children in all the wards, dressed with Santa Claus hats, acting as elves and distributing sponsored presents to the ward patients.

Other leadership positions included Project Make It Happen! Director for Lighthouse School for Blind Children, Project Inspire Chief Editor for Heartfelt, Project ELY (Fun-Loving Youths) Director for KKH Health Endowment Fund and Project Runway to Hope # Beautiful Director benefitting ARC Children's Centre for children with cancer.

He also took part in various community projects including Muscular Dystrophy Association of Singapore (MDAS) Camp Discovery, Club Rainbow Christmas Craft Workshop: Project GIVE (2012-2015) benefitting KKH paediatric patients, Project LOL (Laugh Out Loud) benefitting MINDS (Movement for the Intellectually Disabled of Singapore) and Easter Egg Hunt at Ang Mo Kio Community Centre for underprivileged children.

NUS has recognized him for his dedication in getting involved in various meaningful projects such as those mentioned above, attending workshops, courses and providing service to NUS as shown by his winning various Medical Faculty Awards such as Outstanding Contribution in Community Service in 2013,

Outstanding Contribution in Leadership in 2015, and Medical Society Colours Award in 2013 and 2014.

He is also a prolific book writer and despite his busy schedule as a medical student and now even more busy as a full-time doctor, has penned numerous books which have been published. The books are dedicated to children, usually in palliative care, and their families, and a few are anthologies of poems on the same topic.

Dr Pravin possesses a special gift in this area as this collection of poems shows.

Professor Tan Cheng Lim,
Emeritus Consultant,
Paediatric Haematology & Oncology Service,
KKH Women's & Children's Hospital

Letter to Readers

Do they have telephones up in heaven is an anthology written during my first year as a paediatric resident in KKH Women's & Children's Hospital, culminating an ocean of experiences ranging from the Oncology Service to High Dependency where I experienced different levels within the spectrum of Children's Complex Care and Palliative Care.

This line from the poem 'Gone Too Soon' became the title of the collection because when I read it out at the annual Star PALS (Paediatric Home Care Service for children with life-threatening or life-limiting illnesses), this line struck a particular chord with the families that evening and resonated with them. It made me wonder too how much easier life would be if we just had that connection with our loved ones in heaven. I lost my Dad when I was fifteen and till today, I wish I could make that call no matter how much it would cost. I can only imagine what these parents who lost their children would wish for.

Children's Complex Care and Palliative Care is an upcoming and a new field. While children with complex conditions are living longer, they are a special group of children who need dedicated, special care tailored according to their individual needs. Many of these families have difficulties making ends meet in their quotidian lives and living becomes a struggle. With support systems built in the hospitals and now in communities, these families are getting more help that they deserve. Children with life-threatening and life-limiting illnesses are more relevant in today's world because the emphasis has shifted to keeping them comfortable in their final days. After all, happiness is what a child wants to the end.

Little victories in these fields were little milestones in my journey. From asking the children in the oncology ward to write letters to share their feelings to convincing my first family to create an Advanced Care Plan (ACP) for their child were little moments which I write about and share in this book. I had the opportunity to record my first charity single 'Face the Fight' with a lovely group of supportive friends for 2 children who sadly passed on after the recording of this song. Their voices were the last legacy they left for us which will always be remembered in their swan song. They are also remembered in this little collection.

I hope you will enjoy this collection as much as I did. It is a special collection because each poem has a little story behind it and when I wrote the poems, I remembered the children and their families' journeys as well as how far they have come. I am so proud of all you.

To the children who have passed on, I miss you and think about you, keeping your memories alive in these books and poems. To the children who are fighting every day for a better tomorrow, I hope you know we are all right here by your side, fighting alongside you hoping for a better tomorrow.

Peace and Love always,
Prince P. xx.

Dedicated to all the children & their families - past, present & future,
There is no mountain you can't climb, no river you can't cross & no hurdle you can't overcome,
Have faith.

Gone too soon

Life's clock has struck twelve
The precious time I had with you
Is now over all too soon
Why did you have to leave
Before I could say how much
I love you sweetheart

Is there anything I can give
To make you stay a while more
All those times the doctors said
You have a day more to live
I would hear you had a year longer
No parent wants to let go of a child,
A child like you my angel

Do they have telephones up in heaven
Where I can make a long distance call
No matter how much it costs me,
No matter how long it takes to get through,
I'll wait a lifetime for that moment to come true
Your time with us was far too short dear

When I look up into blue skies
I can only imagine the playground
You're running around in, looking back
Every once or twice trying to recall
Mummy, Daddy and little brother
Memories who will fade as the years pass
But you will be in the centre of my heart
Till my time comes to be with you love

Your little brother looks at me
Breaking down in the kitchen
Quietly wondering why you left
Without growing up to say goodbye
I can't even find the words to

Carry on without the sunflower
That I carried everywhere I went
Wearing you on my heart like any
Proud parent only to have lost all
In a day so dark I almost fell so deep
I thought I'd never recover till

Your little brother smiled at me
Pointing at your photo frame
Then I realised that you'd never leave
The home you grew up in,
The hands who fed you,
The roof that raised you,
That you'll forever be here
In mummy's heart, the heaven
Whose doors will always be open
To you my little angel gone too soon

#Inspiration:

It's been a while since Dreamcatcher – I received a message recently from Geraldine, an old friend from Star Pals to pen a poem for Memorial Day, to remember the children who passed on this past year. This will be my third Memorial Day with Star Pals and each one is more special than the previous one. I see old families whose kids I knew before they passed on and who are in a better place now. This poem is dedicated to all the parents and siblings who have lost a child, a sibling or a friend. Sometimes I wish and wonder if heaven had telephones too…

Harmony Heals

When you're lost and weary
With troubles too many on your mind
Don't be afraid my dear friend
Cause I'm right here holding your hand

Sometimes life can put you down
Stand your ground and hold your head high
When you feel like no one's listenin
Close your eyes and say a prayer

Chorus

That's why
We gotta stick together
Help (pause) one another
Through the thick, through the thin

Through rain, through shine
That's why
We gotta lift each other
Carry (pause) one another

Through the good, through the bad
Through happy, through sad
To build a rainbow of love
For you and me

When darkness falls on your heart
Don't let go of what you believe in
Cause tomorrow the sun will rise
And all your battles will be won

If you ever feel forsaken
Remember you're never alone
Friend I'm here beside you
From now till forever more

Bridge

Cause there ain't no rainbows
Without no rain
All your dreams are bigger than your pains
Look at you, how far you've come

No turning back, march on all the way
Your future is right in front of you

#Inspiration:

I wrote this song for a charity single to raise funds and awareness for Star Pals, the Paediatric Palliative Care Service I closely work with. We had the lovely opportunity to record the song with a group of dedicated friends and a couple of paediatric palliative patients whom we sadly lost soon after. It was their swan song and this will always serve as a memory for their families who still remain grateful to this day. It was my greatest honour to have met these children, one of whom heard the final version of the song which I played for her when I visited her in her final days. She approved the song and was thrilled to hear her part in the song. It made her day and her mother's day too. These were the original lyrics I wrote and I felt this compilation would not be complete without *Harmony Heals*.

Midnight Sun

In the middle of an old town
A cathedral stands tall and still
From the break of dawn
Till the twilight sun
The cathedral I first brought you
To be baptised
The cathedral I now bring you
To lay you down to sleep

I remember running with you
In fields of purple flowers
For miles on end, we'd run
Through the ice-cold winds
As fast as our feet could take us
Till one day you stopped running

I asked you what's wrong
And you said you couldn't breathe
The air that once filled you with
The spring of life was now being taken
Away cause of a growth they found
Deep in your lungs

The growth that took us back to
The cathedral we once thought
Would shelter us from life's storms
Bringing us down to our knees
To believe again in a miracle
To cure you of this growth
That lit up all over your PET scan

There were days when we'd sit
On the bed of purple flowers
Watching the sun set behind
The majestic cathedral that
Somehow stood its ground
Through the age of time
Just like the love between
Mother and child

Wheeling you across
The old fields we used to race
Felt different, overcast by
A sadness I saw in your little eyes
The eyes that finally closed one night
When it rained so hard
I knew the heavens were protesting
Against the loss of another child

Here I stand
Facing the cathedral
I raised you in
The very cathedral
I had hoped to see your wedding
And your child's baptism,
Wondering how fate
Had its twists and turns
Something I couldn't explain
Like how the sun didn't set till midnight
On the day you passed
A sign that you were still standing here
Right next to me by the cathedral
Where you were baptised eight years ago

#Inspiration:

Midnight sun is a special phenomenon that takes place in the Arctic circle. Nature and spirituality are closely linked and a recurring theme in my poems. When I wrote this poem, I imagined a mother and child playing in the fields outside a stand-alone cathedral in a rustic old town. And these spiritual settings bear a spiritual significance for the old town folk where they celebrate life and death. Written in a simple, quiet and honest tone, this poem aims to remind all that a child's passing is a sad event but it is a miracle because they become angels and they will return to a place where the sun never sets.

Mommy when will you get better?

Missed a few days of school
Last week cause Mummy had a fever and flu
Again this time worse than before
They told me to bring her to the hospital
Pushing her wheelchair, I grabbed a cab
And in no time, we were back
Right where we started from

The oncology sisters welcomed me
With a familiar smile of sadness
Feeling all too sorry that
I had to grow up all so soon
Learning complicated terms
Like chemo-ther-apy and neutro-penia

I felt it rain again inside
Seeing Mummy toss and turn
Vomiting through the cycles
I just wish I could wish her well
With one spell – why can't it be that easy

I learnt nothing was easy
The journey with Mummy
Was rough especially when
Cancer hitches a free ride
It's like having an unwanted passenger
You just can't get rid of

I wish Mummy would get well soon
I wish her white cell counts would fall
I wish she could grow her hair again
And not feel ashamed when we go out
I wish life could be kinder to us

It's been a while since
We had ice-cream
Since I'd been invited
To a friend's birthday
I've learnt to always wear masks though
And to practice hand hygiene in the wards
Something I can be proud of I guess?

Sometimes when it gets late in the wards
I sit down alone next to the vending machine
Doing my homework sometimes falling asleep
Only to be woken up by a kind nurse or two
Asking me if I wanted a blanket to stay warm

Rough times will pass someday
Here's to bright and blue skies
For Mummy and me
And for all the children
Tiny symbols of strength

#Inspiration:

I was on call in SGH (Singapore General Hospital) late one
night running the haematology-oncology wards when I saw
this young kid sitting by the vending machine with a mask
falling asleep with a cup of milo in his hands. I saw him
the next morning in the wards and realised his mother had
been admitted. On yet another instance, I saw a young kid
wheeling his wheelchair-bound mother out of the wards
to the canteen downstairs. I admire these young spirits –
staying so positive while assuming such huge responsibilities
at a young age. It took me back to my younger days when
I had to grow up so quickly. Having a sick parent changes
the perspective children view the world. It teaches to be
more positive and that's what I realised through camps I
have taken part in involving kids with parents who have

chronic or terminal illnesses. There is a role reversal which builds character and imbues special virtues in all of them. This poem is an homage to them and a prayer that all their parents will get better soon someday – hang in there, stay strong.

A Song for You

Singing sunflowers
Stand in December rains
Heralding the loss
Of heaven's new saint
Will she ever return one day
To walk with me
Along life's lonely lane

I never wished
This was your swan song
But that's all I have now
To keep me strong
Your words, your will
To comfort me all nightlong

Hope you will be
Happy and free
Hope we'll meet
One day
Where the sun
Meets the sea
Tell me why
It had to be
Me singing this now
When you were
Always the one
With the dreams
To be a star

Come back home
To our doorstep
I am here waiting
To hear your

Little footsteps
For just one,
One more dawn

You know Mama
Had big dreams for you
Bigger than you and me
Sorry you had to leave
All too soon, leaving me
Empty-handed, short of
A soul to hold on to,
A story to tell,
A song to sing for you

Now that you are gone.

#Inspiration:

This poem is dedicated to the most talented singer I have known. Shova was a 9 year old girl who loved Britney Spears and her dream was to sing for Britney Spears. She was a renown singer in her native country and she kindly agreed to sing on my charity single. Sadly, it would be her last recording as she passed on this New Year's Eve. I remember coming home to the news after being on call the night before. It broke my heart to smithereens. We sang together during the recording session and she knew every lyric of the Britney songs. Her mum was a gem too. Her mum did her hair, her make-up and she was ever so decorated for the occasion. I only wished she had listened to the final mix of the charity single before she had moved on, all too soon my dear, all too soon.

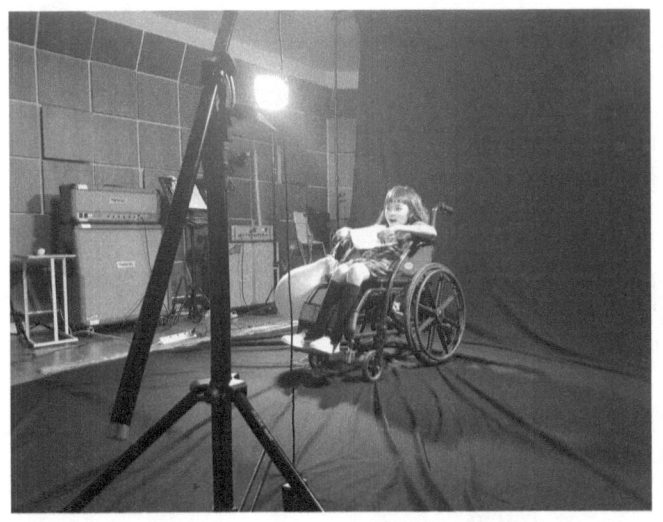

Shova* is a 9 year old Indonesia girl who had Ewing's sarcoma, a rare bone tumour. She loved singing and her mum told us she would always perform in renowned shopping centres in Indonesia to showcase her beautiful voice. It was her dream to sing and she has recorded a music video on Youtube as well singing a song by Britney Spears, her all-time favourite singer. When I met her, I remember she had a beautiful voice and she was dressed up like a beautiful diva, ready to sing and belt her heart out. Her voice can be heard at the outro of our charity single. She was thrilled to record in the studio. Sadly, on New Year's Eve, I received news that she passed on in her hometown surrounded by family and friends. The charity single we recorded with her will serve as her legacy and the poem 'A Song for You' is dedicated especially to Shova and her wonderful family. We miss you and your lovely voice so much. Rest in peace dear princess, we will always remember you love for life and for music.

(*real names of patients have been changed to maintain confidentiality)

In Time for Tomorrow

When I lost you
It felt like forever
But I promise
We'll see each other again
I don't know when
But I can tell you this
I'll be waiting for
That special day

Go ahead now
Don't wait for me
You'll make new friends
But please don't forget me

The night before you passed
My sleep was taken by
A little angel's whisper to me
That you'd breathe your last
In this world
And your first in a new one

Where you'll live a life
Filled with happiness forever
With friends you knew once
Now angels like you up above

In a land of tooth fairies and sandmen
Where fairy tales come to live with
Fairy godmothers, princesses and castles
You'll finally get to wear your tiara
And ride the little pony I could never
Afford to give you back home
When you were tubed and sedated

As we walked the last mile together
Back in the hospital room before
Your flight came to pick you up
I saw your smile so beautiful
I knew Mummy had done
All I could to make you happy

Now it's your time angel
Your time to fly to a place
Where time never passes
So it'll feel like tomorrow
When my flight arrives
When I see you again
In time for tomorrow

#Inspiration:

A Day After Tomorrow was written after the passing of Illy, a precious little angel who also sang on my charity single. She was blind but she learnt the words to the song by listening and she picked up the words so quickly it was remarkable. She was intelligent beyond words and she herself told me she wanted to be a Maths Teacher. In her final days, my friend Camellia and I visited her and her family in the hospital and she listened to the charity single. She loved it and getting her stamp of approval was so important for me. She had a loving family and was always surrounded by love. My wish for her as she moves on is to always have love around her and to remember that she will always be loved.

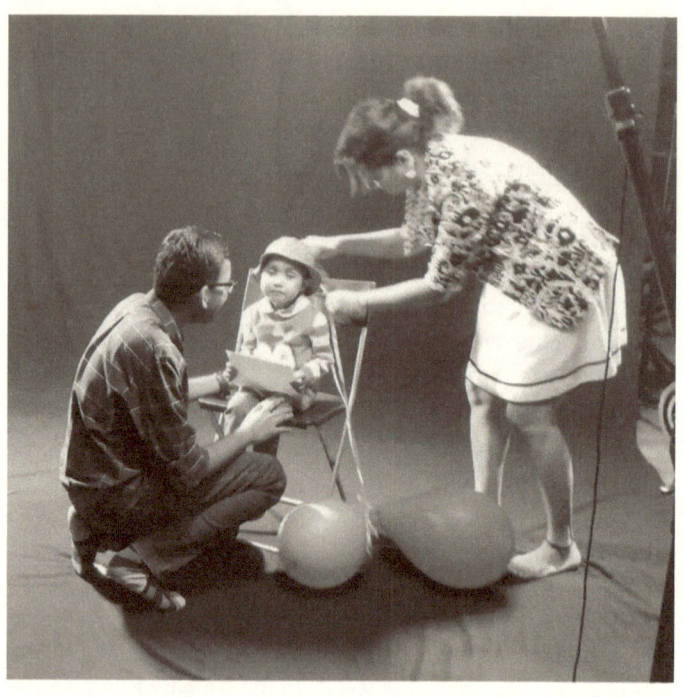

When I first met Illy*, I remember her smile which was most captivating. She was vibrant and her mum she dreamt of being a tuition teacher in Mathematics which was extraordinary. She was a 11 year-old Malay girl who suffered from a rare mitochondrial disorder known as Kearns-Sayre syndrome which affects many organs in the body, especially the eyes. She was blind and she learnt the lyrics to our song by listening to the harmony. Hence, our project title Harmony Heals – because harmonies and melodies can do wonders. When she was ill, we visited her in the hospital where she heard the charity single for the first time. She was thrilled! It was heart-breaking when we received the news that she had passed on soon after Shova. She inspired all of us with her resilience and her hopeful spirits. She did not allow her illness or blindness to keep her from what she loved doing. She lived life to the fullest and that was her message to humanity. We miss you dearly Illy.

(*real names of patients have been changed to maintain confidentiality)

At the Hairdresser's

Usual Sunday routine
At the hairdresser's
Sitting in front of the mirror
I see myself for the first time
Since I lost you.

The radio plays
The good old tunes
We used to dance to together
In your room
Where we played pretend
Running wild in our imagination
Which knows no boundaries

Macy worked through my hair
That had grown long and heavy
With a sorrow I carried
Since the day I laid you to rest

It took me so much courage
To return to the hairdresser's
Where I thought I'd give myself
A little break from life
Only to realise
The old salon brought back
Memories too many.

#Inspiration:

The words to this poem came to my mind while I was cutting my hair at the salon during the weekend off during my oncology posting. I felt a flurry of tears and a sadness come over me as I remembered all the children and their stories.

I then realised that the hairdresser's is a powerful place and the mirror we look into helps us look back in time, taking us back to the past. I came home immediately and wrote this that night itself.

Portrait

Picture-perfect portraits
Hanging on my room wall
Sitting down on my desk
Looking back at me
All I see
Are me and you
Frozen in time

You took my life with you
When you said goodbye
Leaving me with
All these photo albums
Capturing our once-upon-a-time

If only pictures
Could come alive
And tell our story
Over and over again
I'd listen like a little kid
Till the end of time

Talking to your smile
I beg and plead
For your return
Cause I miss you
Time and time again.

Before I sleep
Every night
I remember
To kiss your picture
Goodnight
Like you never left.

#Inspiration:

Photos are powerful. During my oncology rotation, I noticed that parents ensured that the beds of patients were surrounded by photos of their favourite people and moments in time which I thought was very special and endearing. These photos served to remind the children and their families that there was still happiness amidst such difficult times. They are a source of encouragement and give them the will to go through each day even after their children have passed on in some instances. A photo is a special memory – just like every photo in my paediatric palliative care and complex care anthologies which still brings back a flood of memories each time I flip through the pages.

Spaceship

7am rise and shine
Time to reach for the skies
On your little spaceship
Around all the planets
Meeting your alien friends
Who are always waiting
For your special arrival

The big big universe
Can't wait for you
To see your sweet smile
From a million milky ways away
Lighting up all the galaxies
Near and far

First put on your space suit
Then you take a deep breath
And off you go to outer space
Overflowing with timeless love
Floating amidst the stars in the sky
Cause you are after all, one of them,
The brightest one of them all

When you fly so often,
I worry someday you'll fly away
On a one way rocket to space
And leave me behind.

But I know
We'll be together again
Cause you promised
To take Mummy and Daddy with you
We're always ready

With our luggage of love
To join you in your universe
Someday my love

#Inspiration:

As a junior resident, I was in-charge of a lovely girl who had a pontine glioma and was undergoing palliative radiotherapy. Her mum told me she would eagerly wait for me every morning and called me the 'bum bum' doctor because I would always inspect her bottom to ensure there were no new abscesses or tears as she was immunosuppressed and had background chronic constipation. It was the first time I heard her laugh when her mum told me her nickname for me. She was a special child because every morning, after receiving her radiotherapy, her mother would tell me she was back from space and we would check her mouth and bottom for any 'aliens'. She would chuckle whenever we spoke about space and the friends she made there (ie. the radiotherapists). I will never forget her – she was one of the sweetest angels I knew and it was great journeying with her to and fro from space on her little spaceship.

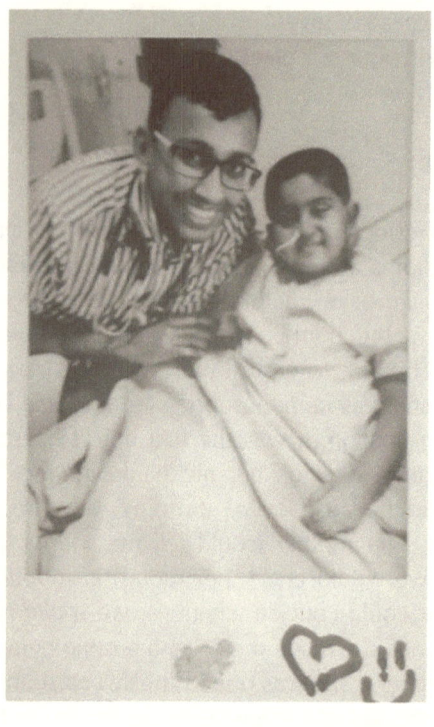

I remember racing through rounds from patient to patient in the busy oncology wards as a first year paediatric resident. One day, I came to review Mimi*, a little girl with pontine glioma, after she returned from radiotherapy and a new friendship blossomed between us. She only allowed me to take her bloods and examine her. I had the privilege of also making a special friend. As the month went by, we became good buddies and I earned myself a place in Mimi's hall of fame (she has this little book in which she collects polaroid pictures of all her favourite friends). Mimi always brought a smile to me, just like many of the patients in the oncology wards. I was always grateful for Mimi and her family. There was a ray of hope in them that shone through their smiles and an undeniable optimism even through difficult times.

Stay strong and hang in there Mimi, am right here with you all the way every day.

(*real names of patients have been changed to maintain confidentiality)

Little Genius

You think I'm small
But I know much more
Than my years
Cancer makes us
Grow up all too fast
Missing out on
What my friends call
'A childhood'

Shuttling in and out
Of treatment rooms and ICUs*
Every day is a blessing
And God I'm not asking for much
Just give me
A couple more ice-cream dates
With my family

I heard heaven
Has an endless supply
Of free-flow candy
But why can't I bring
My family with me?
Why must I leave
Before I get my chance
To grow up?

Every day,
I write in my little notebook,
That I am one step closer
To the finish line

(Trust me I know my treatments,
I used to teach the junior doctors,

Cause I've been through the side effects,
I live to tell)

I hope someday
I'll live to win
The Nobel Prize
For being the littlest kid
To survive the biggest cancer
Wish me luck!

#Inspiration:

My registrar and I were rounding the oncology ward for the first time and we met the smartest kid on Earth. Even though he had leukaemia, he was extremely positive and a child genius. He educated my registrar and I on his treatment regimen and we called him Professor from that day. The following day, he stopped us during our morning rounds and wanted to tell a short story on how too much of anything is not good for us. He did this after I told him to read more during his inpatient stay (instead of always playing games – he actually took my advice!). It was endearing to see how the entire team stopped to listen amidst busy ward rounds and this made me realise that the oncology ward was extremely special because these were no ordinary patients. They were superheroes in their own right. After he told his story, I thought to myself 'this kid was destined for great things'.

*ICUs – intensive care units

Broken Ballad

Now I no longer
Recognise the girl
In the bathroom mirror
No hair on a head
No meat on bone
No soul in a body

With a broken heart
I trudge on trying
To find my courage
On a quest to restore
My spirits, my will to live

How can a singer survive
Without a voice
How can a song come about
Without a melody
How can a ballad be born
Without the beauty of its melody
How can I go on
With a bag of burdens
Without an answer to my illness

One day
I looked out the window
And I saw a young girl
With little hair like me
Thin as me,
Riding a bicycle
And laughing
The way I used to
When I was her age
Before illness took over

When I saw her
I asked myself
Why did I let my cancer
Conquer me
Her smile alone
Was her personal statement
To say
I can fight cancer,
I can beat this
Cause I am better than this
I will and am going to win.

#Inspiration:

I met many teenage girls during my oncology posting, one of whom I drew blood from and initially she was afraid but when I took blood from her, I cracked a few jokes to distract her and she started laughing that even the nurses were pleasantly surprised. She told me the school she went to and her hopes and dreams to get better. I listened patiently even after drawing her blood and told her to hang in there because she was doing well even after having undergone surgery for her ovarian tumour. It is difficult for teenagers to go through the treatment due to the side effects which also affect their body image but it always help when we normalise everything around them to make them feel less isolated and helps them to become integrated back into their routine once again, slowly but surely.

Short Story

Using my tears
As ink to pen
My story
That is gonna end
Shortly.

No second chance
For the cancer
So rare
Feels like I'm
Fighting a losing battle
All alone.

Just when I thought
I was wading
In shallow waters of victory,
I soon realised
The waters became deeper
And soon enough
I was drowning,
Struggling to survive

Why am I chasing a fantasy
When I knew the end
Right at the beginning of the
Chapter on cancer
That it would end painfully
With no happily ever-after

I imagined every story
Will have a beginning,
A middle and an end
But mine seems

To have ended prematurely
With a cancer
So fierce that it
Just won't let me finish
Writing the complete
Story of my life.

God only knows.

#Inspiration:

This was written after hearing about a rapid deterioration of an oncology patient who passed on in the intensive care unit after his parents decided to withdraw support as he was on a heart-lung machine and his parents did not wish to prolong his agony. I could only imagine how he must have fought for his life as he was doing well on his treatment prior to this acute episode. Sometimes I feel that is unfair for children to have cancer and if I had a wish, I wish for all of them to be cured with the wave of a wand so that they can write their story seamlessly without interruption.

Recovery

All too many relapses
I almost gave up
When each count was
Worse than the last

Then one day
When I walked into
The clinic I had become
All too familiar with
I saw a smile on my oncologist's face
She was beaming like
The sun on an eclipse
A rare phenomenon

You're in remission
Those three words
Felt better than
Will you marry me or
I love you
I heard in those feel-good romance films
I loved watching
Only this time
It was real,
I *was* getting better

I was healing
In my heart
Having come to terms
With my enemy cancer
But now that I was making a comeback
I felt the colours coming back to my grey world
I felt larger than the Titanic,
I felt stronger than Hercules

But above all
I felt like a twelve year-old kid again

#**Inspiration:**

Parents and the oncologists placed much emphasis on the word 'remission' because that was the hope they had for all their patients as oncologists always treat their patients with a curative intent. When children are in remission, it means that their cancer has gone away which indirectly means that they have a chance at life, a chance at starting over once again.

Praying for A Cure/Bring Me Back

Stop all the treatment

Pin-drop silence fell in the family room
The oncologist was speechless
My heart skipped a beat
For a moment
I thought I heard wrong

Daddy had tears in his eyes
And Momma was quietly
Staring at the floor
I didn't know what to say
They had finally listened to me

As you wish

The consequences came soon after
With the oncologist painting
A picture so grim I felt
He was all ready to lay a wreathe
On my grave with the words
'I told you so'

But I believe
That life has a reason
For the decisions we make
We weren't giving up
Or at least I wasn't

I needed some time
To breathe again,
To see my brother,
To play with my puppy again,

To draw, to colour
All I had lost to my intensive treatment
All I wanted was to be treated
Not like an oncology patient
But like a kid again
A kid with dreams
Bigger than her cancer.

#Inspiration:

I vividly recall a family conference I sat in to assist with documentation. The parents opened the meeting with the shocking statement that they were considering not to continue with treatment. Jaws dropped in the room. The oncologist could not believe his ears. But I saw a sense of desperation and sadness in the eyes of these parents – they were not being selfish with their decision but it was made on behalf of their daughter because they genuinely wanted a better quality of life for her. Their lives were shaken by their daughter's newly diagnosed nasopharyngeal cancer (she was one in ten million kids to be affected by it and it all started out as a benign headache). I knew them from the general ward I was in when she was first diagnosed.

After the family conference, they spoke to me and said they loved their daughter more than anything but they wanted her to still lead a normal life and felt she needed a break from all the stress. I remember speaking to their daughter and telling her to never lose her optimism. She promised she would always hang on till the end.

Cancer is a difficult diagnosis for everyone, from the child, to the parents to the oncologist but at the end of the day, we have to respect what the child wants because happiness is most pivotal for any child.

I believe a happy child will always heal better.

How I feel throughout this whole journey

so here's how i feel throughout this whole journey. when the doctor told me that he was in charge of cancer patients and that i was one of them, i had no reaction at all. honestly, i didn't know what to feel. i just listened and accepted my fate because i couldn't do anything about it. i had to go through two surgeries. the first time i went to the operating theatre, i tried to hold back my tears so as to not let my dad know that i was scared but somehow tears managed to roll down my cheeks. the second time i went there, i didn't cry because i was used to it but after my surgery, i cried my heart out when i felt something inside my body. it made me uncomfortable that i felt like ripping it off. fortunately, i slowly get used to it after that. as time goes by, i get to know more about my cancer also known as nasopharyngeal cancer. i actually searched up online at home when i was discharged on what it is and the symptoms . it then made me realised that i had this cancer a long time ago. i felt frustrated with myself for not knowing about this earlier and got it treated earlier and i also felt frustrated that i didn't take care of my body well enough .

when my parents told me that i had to went through three cycles of chemotherapy, i had no idea what it was. i searched it up online again and it was actually a treatment that uses drugs for cancer. i felt relieved to know i was still able to be cured. i thought that there wasn't any hope for me but thankfully, god showed me a way. the first time i went for chemotherapy, it was in an isolation room. little do i know that they used a needle to poke into my portacath and it hurts so badly that i cried while clutching my mum's hand tightly. the first round of chemo was so bad that i was so weak till i had no mood at all to play my phone or even watch the tv or even talked or walked. i had no appetite at all as i always felt nausea for that five to six days. of course, i lost a lot of weight but that was also because of the chemo.

the second round of chemo was slightly better although i still felt nausea and weak. this time around, my mum wasn't there for me to hold her hand when they insert the needle into my portacath so i had to hold my caretaker's hand. however, it felt different . i was feeling even more afraid without my mum. i started to lose hair in the process which i didn't care about at first. when i went home, my mum decided to cut my hair short since it was much more easier. i looked at the mirror and i had so much thoughts going on in my mind. for example, i looked ugly, i looked like a boy, how am i going to go around in public, how was i supposed to face my friends and whether my hair was growing to grow back in time if i ever get to go back to school.

as usually, i started to accept it and went for my third round of chemo. it felt better compared to the previous chemos. it went smoothly for me   so after i went home. my parents reassured me that i was going to be okay and i will recover. i know myself too that i'll recover no matter what. the social worker did came to talk about radiotherapy to me and even i don't like it cause it makes me scared to know i'll be wearing a tight mask. regardless of how there are holes on the mask, i'm afraid of small spaces and i get scared that i might even have trouble breathing. so for now, we'll all wait for the results of the pet scan and mri scan. i'm praying so hard that the cancer will be gone or at least shrink. i even made a things to do list when i recover and even some goals. my goal was that i am allowed to free myself of fear from dying and that i will be positive that i am alive and i will recover no matter what.

my things to do were simple honestly. i wanted to get a cat after i recover since i've longed wished for a cat. i also wanted to spend time with my family and relatives and even be more expressful around them so that they would know how i really feel and it'll help to make us even closer. next, i wanted to go to school and meet my friends and we'll graduate together that will make my parents proud. that's all i ever wanted to do after i recover. i feel positive nowadays knowing the end of curing this cancer is going to end soon .

truthfully, there were times where i felt like giving up and just accepting my fate to die. however, i thought of my loved ones who had been giving me moral encouragement since day one. i wanted to make them happy and proud of me for going through those painful needles, chemo and all to recover so i stayed strong and hold on despite how tough and painful it was.

to end of, i kind of have a message for all the patients who have cancer. just want to say that it's okay to cry regardless of how old you are and that you should hold on and stay strong because it's never the end for you. there's this life motto that i held onto since day one of chemotherapy. it's that you either stay or you move on. you either stay to fret over it and get all frustrated, sad which you'll end up ruining your own happiness or you either move on, be positive and strong and of course build up your own happiness which will make your loved ones around you happy as well. so this is the end of my story of how i feel throughout this whole journey.

Thank you :)

An Open Letter

Cancer is a tough business eh? It comes suddenly and quickly catching us off guard. It elicits a roller coaster of emotions in us and the ones we love - sadness, anger, questions such as "why me?" which we all feel and ask cause we are, after all, human.

Now that cancer has happened - what next? Cure - well we always hope for the best. More importantly it is the journey you rightfully pointed out to the hopeful cure we embark on with family and loved ones by our side (like your marvellous mum and dazzling dad who work so hard for your happiness which is the common goal we are all working towards eh?). Life has its twists and turns but tests such as these break and rebuild us over and over again till we finally beat the odds and win.

Is it ok to cry? Like you mentioned, sure as the sun will rise. (If you remember the Disney movie this quote comes from). I cry too - it's ok to let the tears pour and let the words flow out from your heart like you did in the honest and lovely essay you wrote to me.

→

→ It helps us to have a healthy heart (and mind) and that's most important cause it's the heart that keeps us going when the going gets tough. I'm so proud of you (and your family) cause what you're doing and showing now is that You Are A Survivor and keep at it! You're a good role model for all teenagers cause of your determination, faith and positive mindset ☺

★ GOOD LUCK
★ and all
☘ THE BEST ★

My unducking words are to never give up the fight. Cancer may seem bigger than us at times but let me just say win or lose, your fighting spirit has already won our hearts.

All that's left now is for you to always believe in you.

Get well soon and take care ☺
                                        from sincerely,
                                          Dr Pravin . Ward 76.

The children's oncology ward has a wide spectrum of children. Some children accepted their diagnosis. Some were angry. Some were in denial while others were too young to understand. I met this young girl whom I followed from the general paediatrics ward. Alicia* was upset with her diagnosis that had turned her world upside down and when I met her, she silently looked at me, wondering why she had cancer, all the more a rare cancer. I told her to write out her feelings and pour her emotions into a letter to her oncologist. The next time I saw her, she gave me the letter and I wrote her a reply on a huge 'Get Well Soon' card the next day. This letter was very special to me because she wanted me to share her journey and what she learnt along the way with other children in a similar situation. She wanted them to know that they are not alone and that there is someone out there going through something similar. I commended her bravery to write and share her story. These are little heroes in our daily lives who deserve to be celebrated.

(*real names of patients have been changed to maintain confidentiality)

Today I'll cry my last

Today I'll cry my last
When they wheeled you out
Covered in a sheet so still
I felt I breathed my last with you

Come back please
No scream of desperation
Through these doors
Could restore life back
In the child I raised
For so long
Whom I loved so hard

Has anyone heard
A mother's cry for her child
Whom she so unwillingly outlived
With so many questions to God
Left unanswered cause she felt forsaken
In her darkest hour when all she heard
Were whispers of sorry and goodbyes
She knew were in vain

And when the undertaker came
With an expressionless face
He carried your beautiful little self
As I chased after, fighting off tears
Weeping like a widow,
I wished it could have been me inside
Instead of you who was meant to live
Many more years in my place

Won't you please tell me
This was not happening

Nobody told me this was
How was it meant to end
The doctors said you will be saved
But even science did you no good
Why did you leave me all alone

Today I cry my last
Call me bitter, call me insane
But nobody can deny a mother's pain
Nobody bore her child for nine months
Nobody can turn back time
I loved my child like I never loved no other
I loved my child
I loved my child.

#Inspiration:

No one can understand a mother losing her child unless a
parent has undergone the same loss. It is difficult, especially if
the death of a child is sudden and the parents do not expect it.
There is no amount of words that can capture the emotions the
parents undergo and my poem is likely an under representation
of the pain of the loss but I hope that through this poem,
parents can channel their frustration and sadness, knowing
they are not alone and for those who know of such parents –
please remember to empathise with them. It is a tough moment
no one deserves to endure.

I'll Stand By You

Sometimes it feels that
Life has gone so wrong
And all you can do is
Nothing but be strong

I know you wake up
Feeling all so torn
Shredded paper thin
By life's cruel cuts
With nothing left
But hope in between

You sit for hours
Waiting for a fairy
To appear, to grant
Your child a cure
That would save
His life and yours

Then the word comes
There isn't any cure
And all you do is cry
Waiting for an apology
From the author of life
Who put you through
All you've been through

You walk back into
A cold hospital room
Your second home
These past few weeks
To your child,
Your best friend

Who smiles weakly
And you forget
All the anger
Remembering the promise
You made the day he was born

I'll stand by you.

#Inspiration:

This poem was penned to pay tribute to all parents who are experiencing internal turmoil while in high dependency, the oncology ward or even in the intensive care unit. There are days when you feel that you are not hearing the answers you want and there are days when the situation seems stagnant. It is undeniably painful and difficult but I just want you to know that with every crisis, there are good and bad days. Hope is important and even though you may not know it, having a little hope day by day eventually builds a mountain of resilience that becomes unbreakable.

Preordained

When you said
You'd never leave
That's not what happened
When I saw you leave my side
That fateful night
When I lost everything.

When you loved me,
You said we'll be
Together till the end
But when I lost it all
You walked away
Without even a goodbye
Leaving me lonely and forgotten

I walked into the room
But I never walked out
I was trapped inside
The girl I once was
I'm sorry if I didn't tell you
How grateful I was for the love
You generously bestowed upon me
When we were once together

Now I lie around like a forgotten toy
Hoping to be remembered again by
The boy who used to play his guitar
In hope that someday I'll wake up
And become the girl he once knew

Though you never looked back
I still remember how it felt
When we held hands and

Loved each other in a moment
So perfect I can still remember
What true love feels like

I may have lost you
But now I have the nurses
And the doctors and my family
Who still love me

Fate plays cruel games
Taking your love away
But it has its ways of
Returning your love
In this new family I've found
In the little oncology ward
I now call home.

#Inspiration:

I was post-call one morning and covering the oncology ward. I reached a bit earlier to start my pre-rounds. The oncology nurse was handing over the night events and she stopped abruptly at one of our patients who was a teenage girl with leukaemia. This teenage girl had a bleed in the brain secondary to a rare side effect of a chemotherapy agent she was previously on. The nurse recounted how this girl's fiancé used to visit her every day and sing songs to her by her bedside even after she became paralysed but after a while, he stopped coming. The nurse paused before smiling to say that they were now her new family. One can only imagine how the girl must have felt when her fiancé stopped visiting. The power of love can never be underestimated. But we were grateful that she was in a ward where she was dearly loved and where everyone cared for her.

Battle Cry (for the broken-hearted)

Love heals bullet wounds
No matter how many shots
Life takes at us, the greatest
One being childhood cancer
We will make it through

Rise above our troubles
A spirit stronger
No sorrow can defeat
A bullet-proof heart
Defying odds like never before

Call on the elements, call on the universe
Grant us the power to define
The course of tomorrow
May it bring with stories of victories
Of the unbreakable human spirit

What place is there for a tear
Cried too long for it only wastes
A precious moment that could be
Spent praying for courage and hope
To brave the challenges of the
Vast unknown only the heavens control
And only destiny can bring in our favour

Only the deserving are given tests
So tough that there are breaking points
But who ever promised this life would be easy
Yet the force of unwavering faith can never be questioned
And even mountains will bow to heroes who fight till the end
And their names will be engraved in histories rewritten
With every battle won yesterday, today and tomorrow.

#Inspiration:

Having rotated through paediatric haematology/oncology and high dependency, parents receive the most heart-breaking news about their children. May this serve as the battle cry many will remember through their struggles, their journeys and through their most trying moments. When the going gets tough, the tough gets going. Don't ever give up no matter how difficult it gets because as long as there is faith, hope, trust and love, triumph will prevail. Never question the human spirit – it was built to be what it is meant to be for reasons no mortal can explain. Believe in you. Believe in a better tomorrow. Most of all, believe in the power of humanity.

1962

Pull out the gramophone
And put on the record of 1962
Can we go back to the
Good old days when it was
O so uncomplicated

Back to the wild days
As a teenager when
I was carefree as the wind
When I had Daddy
Dance with me at my prom
And Mommy and my sisters
Sing at my birthdays

No one ever prepared me
To face the reality of today
Here I stand alone by the
Window pane, listening
To the rain's quiet sympathy

Now I lay on the couch
In my worn out blue jeans
Wishing I was Daddy's little girl again
I look around only to find hospital equipment
For my little baby, born a little too early

Spending nights watching saturations
Titrating the oxygen requirements
Worrying about alarms that disallow
Any form of rested sleep, I walk on
To my Momma's room and sit down
Browsing through old photo albums
Of the good old days when life was

All too easy, all too simple,
O somehow I wish I would wake up
And it'll be 1962 all over again.

#Inspiration:

In this poem, a single mother reminisces about her childhood days while struggling to care for her premature child who is oxygen dependent. There are days when I do my ward rounds and imagine parents when they were children themselves, living a carefree life. Good memories are healthy ways of keeping spirits up. Happiness is always in the heart of the beholder.

Can't You See

My little porcelain doll
Lost an eye one day
When she could
No longer see clearly
Like other children

Left with an eye,
She started her treatments
Till one day we realised the
Tumour had spread
To the remaining eye

When the news came
I broke down by her side
Picking up the toys on the floor
Trying to hide my tears
from my little doll
who was on the verge of blindness

I wish I could give my sight
To save my child's
The words from the doctor
About treatments to come
Fell on deaf ears

I felt my world darken
I can't imagine yours
Any darker than it already was

God can't you see
This is a child's life
You are toying with
Can't you feel

A mother's pain
To see her child turn blind
Before her very own eyes

Can't you see or have you
Yourself chosen to be blind
To my child's plight.

#Inspiration:

We were waiting for an elective admission for a child with retinoblastoma. We were in the lift when I turned around and noticed a beautiful little girl with a lovely wig with a missing eye. She had lost an eye. When the results about the tumour status for the other eye came back, it was not good. Her mother broke down by the bedside and it was the saddest day ever for all of us, including the oncologist delivering the news. Being a paediatric oncologist is not an easy task but I admire their strength in the fierce pursuit of cure for their patients. When they left the ward, I remembered this beautiful doll holding her little doll and smiling at me. Children have an unbreakable spirit that is special. I can't explain it but they are the reason why I am doing what I am doing. Peace.

Acceptance

The biggest word
For a grieving family
Easily said
But not as easily received

Can you imagine
Your world turned
Upside down when
You hear your child
Has a condition
You struggle to even pronounce

The longer the name,
The more unknowns
About the condition
The more obscure the treatment
The more challenging the voyage

To the layman
Acceptance is a mere term
But if it was your own child
How would you feel about
Hearing your child may not live
As long as you did to see life

When your child looks at you
You can't bear to wonder
What the illness would do
In times to come and
All you can do is to hope,
Hope for the best

Even though that is
On its own as hard as it can be
Acceptance, funny word
You hope never to hear it
Till someday you accept it
For what it's worth.

#Inspiration:

There are always 5 stages of grief – denial, anger, bargaining, depression and acceptance. Acceptance is the hardest and hence it is at the end of the process and even when a family has reached the stage of acceptance, there are still many questions left unanswered and many worries the family may have. It is important for the physician to guide the family through and give them time to accept the diagnosis and understand. Even if they do not accept it eventually, families are only human at the end of the day.

Code Blue

Sometimes you feel the pain is
Too much to bear,
You can't control the tears
And all the little things
From the photos to the places
Remind you all too soon
Of a loss too hard to bear

I used to take for granted
Your first cry, your first smile
When I see other children
I remember,

The times I put you to sleep
With a story and a kiss to
Tell you I will always love you

I look at my phone
No more missed calls
From the hospital
It's been a while now
Since you left,
But I can never forget
The final moments
During the code blue

When my heart stopped
For a moment
But yours stopped forever

I remember the anger
The tears I cried from
Dusk till dawn, my tears

Flooding the heavens
With a sorrow so deep
The angels couldn't help
But cry tears of rain
For a child they once knew

The flowers and condolences
Inundated the funeral parlour

I stood all alone
The world darker than night
So long to the happiness
I once knew
I sat by myself
Knowing there were no more
Goodnights and this was goodbye
For one last time.

#Inspiration:

Whenever a code blue is called in a hospital, it means a patient has no pulse and requires urgent attention for resuscitation. At times, especially in a children's hospital, it is unpredictable and when a child passes on, it is difficult. When a parent loses a child in the most unforeseen of circumstances, it is devastating. No one amount of comfort or consolation can ease the grief. This introspective poem is written in the shoes of a parent who lost her child in a code blue. She remembers and reflects the different phases – from home to hospital. Towards the final lines, she runs out of words to express herself and she sits alone to say her final farewell.

Last Chance, Last Call.

Last Friday I picked
Up the phone
Hoping you'd answer
My call
But all I heard
Was a silence too
Hard to bear.

I wish I had
A longer time
To tell you my dreams
My prayers, my hopes
For you had you lived on.

Answer me one last time,
Please don't you fade away
Cause this is my last chance
To hold you, to dance with you
In those tender loving arms.

How I call and I call
Trying to reach you.
How hard can it be
Even though you just left me
Not too long ago...

#Inspiration:

Like the title of this anthology, a phone call to a loved one is my euphemism of reaching out to those in a world beyond ours. Sometimes when we think of someone who has passed, we speak to them, hoping they would listen and answer in return. It is hard not to hear a reply when we ask questions such as

'Why did you leave me?' or 'Is it fair to me?' Questions we all hope we will find answers to someday. I know it is not easy. I have asked those questions myself too and till today, I have not found the answers. But it is important to not lose faith, to not lose hope and to always trust that someday our phone calls will be answered, someday.

Silence.

Held a candle
For once in my life
Brightly lit
With a photo
And your name
Engraved beneath

A little tear
Streamed down
As I told myself
It was gonna be alright
Though I asked myself
When will I be ok again?

Paid your funeral bills
Cent to cent, till all
I had left was an
Empty pocket, empty home
And an empty heart

Kept the house lights dimmed
And the music ever sombre
In your late honour
Leaving me paralysed from within
As I cried doin' the dishes, wishin'
I'll have someone to cook dinner for
Someone to love, someone to hold

As I blew out the candle
On this faithful August night
I stood by the wooden door
And listened for your footsteps
Hoping you'd return

To remember
The life you once knew

Silence. It fell deep
Through my heart
The same silence
I heard the day you passed
A silence to keep me
Company for the rest of
My days shall be lived in
Silence.

#Inspiration:

I remember when my Dad passed on I felt a deadly silence fall
upon me in the intensive care unit at the cardiology centre.
I felt the tears and I saw the tears from my mother's eyes. It
was the hardest moment. I feel the same silence in my hour
of need when I feel all alone and abandoned wishing my Dad
was here again. When a family loses their child and remembers
on their death anniversary, we always uphold a moment of
silence. Silence is powerful. Silence is solemn. Silence evokes
memories. Silence is respectful. I wrote this poem as a tribute
to silence – often underappreciated but it speaks for itself in
quiet ways.

As long as...

As long as your heart is beating
And your brain is breathing
I know you will live

When I first saw you
You came into my life
As my beautiful daughter
With a smile so radiant
It could light a million towns

Then one day you weren't yourself
Something was not right
We brought you in through those hospital doors
The very doors you never came out off

How we clung so tightly onto a hope
So fragile we almost gave up all too often
Some days you were fighting for your life
On others we were fighting to hold on

The infections kept coming
The chemo didn't seem to be working
The team was trying, we were crying
But then we believed

As long as your heart is beating
And your brain is breathing
I know you will live

When late one night
The doctors came out
With a heavy heart
They said she suffered a
Massive bleed in her brain

Your heart was beating
But your brain stopped breathing
And our hearts stopped beating
That very evening

What ever happened to all our prayers
Well they kept you alive for as long as
They fuelled our hopes for you to live

With a heavy heart we bade you goodbye
Thank you for holding on for as long as you could
For the both of us and for yourself
We believe the heavens have taken control now
Of your destiny so great we will never understand

I will miss you my dearest daughter
There is no greater hurt for a father
To write a sadder tribute but I hope
This will bring comfort to all Daddies
Who lost their daughters because
I want to believe and I hope you do believe
They are in a better place, they may not have
Won their earthly battles, but they have earned
Their rightful place in heaven

I gave up my all, my job, my life
Only to lose my precious daughter
I will never be able to comprehend
Life's mysterious lessons but I believe
That children are little angels who
Come into our lives to change us
To become better human beings

As long as I have a memory of my daughter,
I am grateful for all the times we shared
As father and daughter. I love you dearly.

#Inspiration:

I remember a young girl whom I saw in my last week as a junior resident covering the High Dependency Unit. She was the most beautiful little girl I had seen. Unfortunately she was diagnosed with a very rare form of a lung malignancy. Her odds of survival were limited but she lived for almost a month with ECMO (extracorporeal membrane oxygenation) support. ECMO is a life support machine that aids with the exchange of gases when the body's main organs such as the heart and lungs are unable to do so. I remember seeing this family outside of the intensive care unit a couple of days before she passed and their spirits were shattered. I told them that as long as her heart was beating and her brain was functioning, she had a chance to live. Two days late, I received a text from a fellow colleague who mentioned she had passed. It was devastating and I could not believe it. We were all distraught and I could only imagine what the parents were going through – especially her father who had given up everything to spend every living and breathing moment with her. It was a dark day and my prayer is for all this family who held the fire of hope burning bright till the very end.

Little Heroes

Maybe little heroes
Are never meant to stay
Far too long in a world
Where their powers
Serve little use

God created them
For the greater good
of Mankind to serve
A greater purpose
To teach us a lesson
or two we didn't know

Little heroes
Deserve to soar
Fly high in the sky
No Earthly chains
To bind them
To hold them down

Just when I thought
I lost my little hero
I realised that he is
Using his superpowers
To save a billion more lives
Somehow, someway
I can't understand
Cause God made them
Little heroes special
In a way mere mortals
Like us will never know

Or maybe we were
Simply never meant
To know, and to leave
These little heroes be.

#Inspiration:

When we think of heroes, we think of super natural beings
with supreme powers surpassing Mankind. Children are little
heroes and the passing of a little hero is not easy but like a
phoenix I believe they will rise high and above to serve their
divine purpose. At the end of the day, most importantly, we
all have our little hero who watches above us like a star high
in the sky. We may never understand God's reasons but we
know that God created these little heroes to light up our lives
and our hearts.

Angel's Circus

Yesterday the circus
Came to a little town
Where an angel
Once lived
This angel who
Used to love
All the circus could
Bring with it

From the dancing animals
To the entertaining acts
The angel that sat at
The very first row
Smiling like a star
Each time the ring master
Waved his baton once
Even calling angel to
Take part in an act

Sadly one September
When the circus returned
The acts wondered where
Angel had gone, leaving the
Front row empty of her smile
The animals wondered
And the ring master pondered
Why o why was the angel
Once upon a time an avid fan
Now nowhere to be seen or heard from

Then one night the acrobat
Picked up a note left behind
And it read, Angel left us a while ago

But she always loved the circus
And it held a special place in her heart
Thank you for the magic
That kept her alive all these years
That made her happy in her final days
That left a smile when she left us
For a circus far, far away in a land unknown.

#Inspiration:

Children love the circus. A circus is full of vibrancy, fun and laugher. Children love the animals, the magic and the fantasy surrounding the circus. This poem tells a story of a little angel who used to sit in the front seat for every circus show till one day she passed and moved on with special memories of a circus she'd never forget.

Palliative

I've been hurt, broken, in pain
One too many a time
It's like they don't care
How I feel
Am I just another statistic
Just another failure

Couldn't go wrong
That's what they told me
Now just look at me
All that's left is this
Cancer that's eaten
Most of what was me

Mistakes are human
So are doctors
Chemotherapy works
Radiotherapy works
But all I see are scars and suffering
Of a person I once called me

Can't you see
What you've done
You've made a Frankenstein outta me
I have nowhere to run, nowhere to go
All that's left now is to cry, cry till it hurts
Cry myself to sleep till the pain is gone
Or at least till the morphine kicks in

You branded me palliative
Promising a pain free journey
But all that's earned me
Are tons of looks of sympathy

From passing medical students
And relatives who don't know me

I wish someone would have told me earlier
What I had gotten myself into when I said
All I want to be is comfortable cause
This is the most uncomfortable I've felt
I didn't ask you to feel sorry, I just want
To feel okay, just like I did when I was well

But hey now I'm palliative, guess that's
What I signed up for, an unwarranted sorry sign
Stuck on my door, a sad sorry
From strangers who've lost their sanity
Looking at me like I've lost mine

Why do this to me
When all I asked
Is to be at ease
Give me the morphine
And leave me be
After all my days are numbered
And my end would just mean
Sorry she left so soon

Palliative; people don't get
The true meaning of the word
Making one feel human again
And treating one like a human.

#Inspiration:

This poem was written to reach out to all who misunderstand
what palliative care means. It portrays the misgivings of a
patient depicted as palliative. It is not a death sentence and
it does not mean the end. It simply means helping to make a

patient feel better by alleviating their symptoms such as pain or discomfort. It is often a taboo word in many societies till today, especially in conservative Asian ones but it is important to break taboos and to reinforce the truth of the modern day definition of palliative care – to make one feel human again by treating them with humanity, respect and dignity.

Bricks

I built a life
Out of sand
Right from scratch
To see it grow till
Where it is today

It took a brick
To build a house
Set in stone
Standing tall
Amidst a storm

Sometimes it's hard
To stand tall in a time
When it's so easy
To hide amidst shrubs

Sometimes being left
With barely nothing
Pushes one to reach
For the stars
Cause each star
Earned its place
In the night sky

Bricks can be thrown
At a weak foundation
But don't you crumble
Use those bricks to
Build yourself up again

At the end of the day
When we return to

the very Earth we came from,
The very bricks that built our lives
will be the very bricks that form
The building blocks of tomorrow.

#Inspiration:

This poem was written to remind everyone of the circle of life. We are built from bricks and the bricks we are built from will form the foundation of tomorrow. Life is never easy with its ups and downs. However, it is always important to pick up the bricks hurled at us and use it to construct a better and brighter future. It is always easy to run from our problems but to face them and to learn from them will only make for a better and stronger human being – stronger in spirit and stronger at heart.

A little faith.

I know you have travelled
Long and empty roads
With no end in sight
Carrying a heavy weight
Like a burden on your heart

Do you recall
The hour when the
World seemed darkest
And you were so strong
Shining like a diamond
Bringing light into the lives
Of all around you

You read of great heroes
In children's storybooks
Not realising that you've
Been your child's greatest
Hero in reality

When you feel the
World is against you
And you're alone
To fend for your child
Just remember that
Challenges make survivors
And you're one of them

You may think all
Those tears cried on
All those silent nights
Are in vain but never forget
That the sun will rise

And tomorrow will come
Bringing along with it
A better you

Have faith.

#Inspiration:

There are times on a journey with a palliative child or parent when we feel that the weight of the world is too much to bear. It is a difficult and we are often filled with doubt and apprehension about the certainty of tomorrow. You, my readers, are your greatest heroes. Never forget that. Adversity has chosen you because you are a survivor. When the candles of hope are dim, have a little faith and press on for a better tomorrow. Keep fighting.

I Wish God Knew.

I wish God knew
How much suffering
There is on this Earth
With children living
In unmeasurable poverty
Going hungry without
Food and even a drop
of clean, drinking water

I wish God knew
The answer to help
Eradicate famine and
The waves of drought
And the many illness
That plague the world
Today when even
Medicine cannot heal
The broken-hearted

I wish God knew
The tears of suffering
The pain parents endure
When a child is dying
With no cure in sight
With no peace of mind
And no hope at heart.

I wish God knew
That prayers need answering
Just as the many questions
Of injustice, hate and wrath
That create wars in this world
Which divide humanity,

Which take innocent lives,
Lives that matter.

I wish God knew
That we are grateful
For all the blessings
To be able to live and breathe
Each new day with a purpose
To make the world a better place.

#Inspiration:

This poem is written from a child's perspective. This poem is meant for all in this world who are suffering in a famines, in endless wars or with incurable illnesses. Remember God is watching, listening and waiting for the perfect moment to answer our prayers. We may not understanding the workings of heavenly powers but we can always do our part by being good and doing our little bit for humanity in what small ways we can. Peace and love always.

You Make Me Smile

When skies are grey
Who is my sunshine
No one but you
And your little smile

When the mood is blue
And happiness kept at bay
It all turned around
The moment I saw you

Somedays I feel sad
Wondering how to get through
Another day but then you look at me
And say 'Daddy I love You'

I wish I could say
That every day
The only reason I stay
Content this way
Is because of you

When troubles come my way
Too big for me to solve
I realise I am not fighting alone
Cause you are by my side
With your little sword and shield

It is no easy feat
To achieve what we have
But baby I could not have
Done none of it
Without your smile

Your smile fills my heart
With a million diamonds
Priceless like a shooting star
Bringing my spirit so high
That the sky is the limit for us

You make me smile, yes you do
Not even a thousand suns
Are enough to brighten my day
No amount of words make me
Feel as good, as happy, as peaceful

Daddy loves you my shining star.

#Inspiration:

I wrote this poem after one of my favourite and most memorable experiences in the oncology wards. I was racing through subspecialty rounds when I heard someone call out my name. I stopped and turned around, and lo and behold, I saw my patient and his mum waving at me. It had been several months since I was last in the oncology ward. But this little kid and his family had endured much together and I recall being with them when the little kid was diagnosed. He called me Uncle Papadum* (which was a favourite Indian treat of his) and we promised to have a little papadum party after he got better. He and his mother made me smile so brightly. The little kid had an infectious smile which I caught. This moment inspired me to write an uplifting poem so aptly entitled 'You Make Me Smile' (o yes you do – I hope you're smiling too after reading this :D)

*Papadum is an Indian delicacy that is a cracker often eaten as a side dish.

Lonely this Christmas

Merry Christmas
To little Oliver
This year's Christmas
Is just a little quiet
Without you

Can't bear to open
Another present
Without you around
Remembering all those
Years under the Christmas tree
When you were the angel
Whose smile lit up my world
While you tore open the presents

This year's gonna be
A little different without
My best friend by my side
The holiday just isn't the same
Without your voice to sing
My favourite Christmas carols

I wish I wish
We could have spent
Just one more Christmas
Together, how we used
To wait by the window
For Santa Claus, you
With your milk and cookies
And I falling asleep by your side
Till I'd wake up the next morning
And find you fast asleep on my lap
With the milk and cookies all finished

And a tiny milk moustache on your lips
That I'd gently wipe before carrying
You back to bed with a little kiss
On your glistening forehead

Another year gone, I think
Of your birthdays and Christmases
The two days of the year
You'd be the happiest
I miss you when I think back
Of all the memories that
Even money can't buy back
Come back my Oliver
Won't you please

This Christmas
Feels like another day
An empty one
Without you
Running around
While I decorate
The tree

These twelve days
I'll build Frosty the Snowman
I'll bake a log cake
I'll sing Rudolph the Red Nosed Reindeer
I'll go visit the Homeless Shelter
To spread some festive cheer
I'll hang up the decorations
I'll set up the fireplace
And have a cup of hot chocolate
Like we used to back in the day
Just the two of us

This Christmas
As I say my prayers
I remember to be
Grateful for the
Times we shared
Through the years
You may not be
Here to celebrate with
But I have you
Safe in my heart

Though I feel
All alone this Christmas
I'll remember last Noël
When you said
'Mummy you're my angel
Merry Christmas to you'

Merry Christmas to you too little Oliver.

#Inspiration:
Charles Dickens wrote The Christmas Carol, which thanks to my mum's influence, is now my favourite Christmas movie of all time. In the novel, there is a character known as Tiny Tim who almost instantly earns the love of readers. Hence, little Oliver was created for this poem. I write this poem because no anthology is complete without a Christmas poem, the most special day that brings memories to all those who have lost a loved one. The spirit of Christmas magic in the air brings renewed hope to all for a better year to come. Last but not least, it represents a time for quiet reflection to cherish all those whom we have loved and lost – they are the angels watching over us, always. Peace, Pravin xx.

Happy & Free

I had a bad dream
Last night
Saw myself running
Away from monsters
Monsters too many

One of them carrying
A big, sharp needle
To inject me with
A poison that would
Weaken me till
I fall to my knees
Crying for freedom

Another trapping me
In an ice cold room
Where no one is allowed
No one but an unfriendly robot
That serves me
My meds and meals
Without a tinge of love

I woke up crying
Through many nights
Only to see my Daddy
Standing next to me
Telling me it's okay

Daddy is my superman
And I am batman
Together we will defeat
All the evil monsters
I know we can defeat them
And one day we will all be
Happy and free.

#Inspiration:

I was rounding the oncology ward on a Saturday morning after my call ended. I went to see a patient for whom I had accessed the portacath* the previous day. He was a friendly kid who taught me how to play drums the previous day. Before accessing the portacath, I told him I was going to access the little drum in him. We became friends soon after. He started on his chemotherapy but he had nightmares that night. This made me wonder if chemotherapy causes nightmares in kids. The next day he woke up and while I was examining him, he said I was Batman and he was superman. His father and I broke into a smile. I felt at that moment that kids always believe good will defeat evil and they personify their superheroes who represent elements of good to them. That very child inspired me to write this poem :)

*portacath: this is an implantable central venous access device that serves multiple purposes such as drawing of bloods, administration of medications such as chemotherapy and are often used in patients who require long-term medications.

Spirit

Candle in the darkness
A quiet flame dances
In the silence
Void of melody
Just to the flow
Of a burning reality

A fiery reality
Where hope was extinguished
By the furnace of fear and loss
Of innocent children who did
No harm to deserve the punishment
Of an illness so grave without cure

Closed eyes with tears streaming
Down a worn out creek almost dry
From the drought of sadness
Lasting many days and nights
So long that even heaven
Deserted the broken heart
Leaving it burnt and bitter

Crushed in spirit
Broken down and torn
There is nothing left
In this ravaged town
Where life has been taken
For only sorrow in return

For a moment
The mirror looks back at me
And from a corner I see
A spirit of a child I once knew
Looking back amidst the flames
Of anger, regret and remorse

The candle dies out
As the melting wax
Burns through my emotions
I place your photo back
In your room,
Where your spirit resides,
Where my heart belongs
Right beside you.

#Inspiration:
When someone passes on, their spirit remains. There is anger, denial, hurt, bargaining and sadness. We experience a flurry of emotions and everything seems irrational when we lose someone. I remember waking up in tears during some nights when I lost my Dad. It is hard – you feel alone and no one in the world understands you. I've been through that and I wrote this poem to comfort all families who lost a loved one – be it a child, a parent or a dear friend. There is no remedy. We will go on a journey with time to heal the pain – hopefully and eventually. To be honest, it hurts but it helps to know that someone is watching over you in your hour of need and through all the good times like a guardian angel from afar.

Sacrifice

Beyond the ends of the Earth
Till the end of time
Mummy and Daddy will do
All that it takes
To get you better

We won't surrender
To this grave malady
Not ever so easily
We may not win today
But we still have tomorrow
To conquer together as one

We've given up everything
To give you the best you need
Happiness has no price
And neither does health
As long as you are comfortable
Our hearts are at ease

Studies tell us survival numbers
But you are no study, you're our child
And only God has our story written
And your chapter has just begun
With so many more pages to be written
A tale of survival and resilience
Beyond measure

This highway is long and never-ending
Full of roadblocks, times we had to reverse
When we thought we could drive through
It hasn't been as smooth as we thought
But we will get to the final destination
Together as one no matter how long it takes.

Sacrifice, that's what family is for
It does not matter how much
Time, money, sweat or tears
It may take an eternity but to see
Our child get well will make it
All worth it someday
When we look back at the
Countless sacrifices we made
For our child who is
The future of our tomorrow.

#Inspiration:
In the oncology wards, in high dependency and even in the intensive care units, I have seen parents make the ultimate sacrifices from giving up their jobs, and spending all the time with their children to see them get better. Sacrifice is a big word but with even bigger rewards. It takes energy, time, and it is difficult but it all pays off when children get better and turn around. It becomes worth it. Don't give up even when it seems there are countless obstacles standing in your way. Have faith and keep moving forward – never look back, only forward. One day, it will all pay off – patience.

Winds of Time

I sit here by the window
Flipping through the pages
From your first birthday
To your first day in school
Never thought I'd had to face
Thinking of your last days

As I look out the window pane
I see falling leaves in the autumn rain
And I remember how we used to be
So happy together till the winds of time
Blew us apart, now I can't even see you
All I can do is think of and remember you

I begged the angels for one more day
But they said time was going to run away
With you, leaving me with nothing but a memory
That time will never be able to erase
No matter how many days, months or years pass

I hoped the winds of time
Would bring with them
Winds of change, hoping you'd heal
But you never got better
And the winds stopped blowing for a while…

Always wished time could bring us back
To the good old days when you were here
But I know that time never waits
Even for a day more I'd wished I'd spent with you
Does time not have mercy for a mother's plea
Does time not have a moment to spare for a father's wish
To have a moment more with their beloved child

Sometimes it is so hard to ask so much of time
When time asks haven't we been given enough
When it comes to a child, there is never enough time

Stop chasing the train of time
Those old memories will fade
Say old friends of mine
How can I ever forget your first cry
And your first smile, your first day
And your last everything

Though the winds of time
May have swept us apart
There is no taking away
Moments so special and tender
We shared as family
Blood runs thicker than water, no -
Blood runs thicker than time
Our love for you my precious
Has no limits, has no expiry date
That time can place

I know you are in a place
Where time has ceased to exist
We hope to join you someday
When our time is rightfully up
In a place where the winds of time
Will never ever blow our way again.

#Inspiration:

As a closing poem of this anthology, I felt that Winds of Time
will bring a timeless element to this collection. Juxtaposed
against how time and tide waits for no man in our world, time
ceases to exist in a world where children run free as angels and
never have to worry about a moment gone too soon.

Thank You's

It's been a long journey since the beginning and when I look back at everyone who has been with me through the good times and the bad times, I can't thank you all enough. Thank you to everyone who stuck around and believed in the cause of Children's Palliative Care and Complex Care. I believe there is always hope for a better tomorrow.

Thank you ◊ God for giving me the strength to keep up the fight for these children and to raise awareness of their plight ◊ Mum for the unconditional faith you have in me that has propelled me this far ◊ Dad for the inspiration to always follow my dreams no matter what ◊ my brother for your little tricks that keep me on my toes ◊ my grammy and grampy I love you to bits ◊ Mdm Marjorie Tan for always keeping me in prayer and keeping my fire to write burning bright ◊ my best friend Lu Si Yinn Lulu you're the best ◊ Zi Hao & Tri my pals from way back when! ◊ Hester am proud of you always! ◊ Marianne for your heart of gold ◊ Zoe Tan my plus one gossip queen love ya! ◊ Wei Qi bestie TTM ◊ Katrina bless you kind soul I love our little meals ◊ Dr Judith you are an inspiration, hope to be as great as you someday! Workin on it ◊ Dr Lee Jan Hau stay cool 4ever ◊ Dr Esther you're an angel keep shining ◊ Dr Joel for taking me through my baby steps & a sense of humour ◊ Dr Veena Logarajah for your amazing spirit I appreciate it ◊ Dr Indra for mentoring me through thick & thin ◊ Dr Lynette Goh for sharing the food always! ◊ Dr Lynette Tan cool as a cucumber ◊ Dr Lynette Wee for encouraging me to keep writing ◊ Dr Valerie Ho for guiding me from way back when ◊ Dr Cristelle for believing in the purpose I fight for ◊ Dr Raveen for believing in me ◊ Dr Ng YH for being there ◊ Dr Kumu for the life advice ◊ Dr Oh JY for the reflective moments ◊ Dr Irene for being a gem ◊ Ka$h-fi for the laughs, for the friendship and for the singing moments t'was fun ◊ Anu my sis from another mum ◊ Chris Seow da best in biz ◊ Sheau Yun my dearest friend for looking after me ◊ Devaki for the motivational messages ◊ Lynn you're exceptional ◊ Joselyn you're such a good friend thank you for listening ◊ Heidi you love your patients so much it is amazing ◊ Dr Mya our life stories are so similar, you're an inspiration! ◊ Aimee Teong miss you so much! ◊ J.Loh thank you for being there ◊ Ian Loh my sweet heart through my Mariah moments ◊ Ying Tai my fav bae ◊ Kegs Lim your voice shining bright on my single ◊ Cammie for being there since time immemorial ◊ Nish for being ever so supportive ◊ Prof Tan Cheng Lim for the lovely foreword ◊ Dr Soh for taking me under your wing since I was a student ◊ Prof Tay Sook Muay for being my Saviour on so many occasions this is only possible because of you ◊ the entire KKH Haemotology-Oncology department for being so special and kind to me, you mean so much to me ◊ Logesh keep at it you'll get there ◊ Jessica Ee, Yi Ting my fav sistas y'all rock my world ◊ Si Min rock steady ◊ Yee Nam for being an awesome buddy! ◊ Colin Ng love your designs ◊ Xin Yan & Vivien miss our clinical daze ◊ Kenneth Chin for being the radio rockstar ◊ Huang Peachy you are my favourite Patrick Star ◊ Dr Samantha for being my compass ◊ my Star PALS fam you're the reason I keep believing – thank you Dr Chong, Serene, Geraldine, Kay, Lily & team, I heart you much much ◊ Dr Riza, Dr Maria – I love you much ◊ ma famille à Montréal et MCH vous me manquez bcp ◊ Dr Barathi for the uplifting messages you have no idea how timely they are ◊ Dr Jeanette Goh for the your encouragement alwayskind words ◊ Sara, Hudah & Saidah for your lovely voices ◊ Katherine Leong for your acting & for your friendship ◊ Calida for sharing the passion for our complex care kids ◊ Tingz for the hilarious moments ◊ Prof Chan MY for being a kind mentor ◊ Prof Tan AM for the animated moments ◊ Dr Suresh for being a patient research advisor ◊ Stanley, Nicole, Eugene – my awesome co-HOs, life would be less colourful without you ◊ Prof Chay Oh Moh for always watching over and the little conversations we have along my journey ◊ to all the children & their families who are in this book I miss you & you're always on my mind ◊ to all who are fighting this struggle please know I am right here next to you and don't ever give up for together, we can make it happen ◊

Other Titles by R R Pravin
African Girl (Authorhouse 2008)
When Angels Bleed & Devils Lie (Authorhouse 2009)
Phantom of Keys (Authorhouse 2011)
Eyes of A Broken Warrior & Other Short
Stories (Authorhouse 2013)
The Music Box (People Trends Singapore 2014)
Caught in the Mo(u)rning Rain (Authorhouse 2014)
Dreamcatcher: Le Capteur de Rêves (Authorhouse 2015)
The Good Girl's Guide to Mean Boys (Authorhouse 2016)
Do they have telephones up in heaven? (Authorhouse 2017)

Compilations & Collaborations:
Heartfelt: A Compilation of Short Stories (Yong
Loo Lin School of Medicine 2013)
Timba's Rainbow Journey (English Corner
Publishers Singapore 2013)
Will's Magical Christmas (English Corner
Publishers Singapore 2013)

Website:
Hope You Can Support My Children's Palliative
Care Facebook Page by Liking https://www.
facebook.com/palliativecareforchildren

En chère mémoire de mon père, tu me manques beaucoup

This is not the end, just the end of a brand,
new beginning. Gros bisous xx

www.ingramcontent.com/pod-product-compliance
Lightning Source LLC
Chambersburg PA
CBHW030900180526
45163CB00004B/1649